Photoshop® 2021 Introduction

Quick Reference Guide

Comprehensive video training & instruction manuals available at www.teachucomp.com

TEACHUCOMP, INC.® ...it's all about you

The Photoshop Environment

1. Menu Bar
2. Application Title Bar
3. Tools Panel
4. Open Documents Tabs
5. Options Bar
6. Canvas
7. Active Image/Artboard
8. Panels
9. Status Bar

File Management

Opening and Closing Files

Upon startup, Photoshop displays the "Home" workspace, providing a single, convenient location to create new files or open existing ones.

1. **To open an existing file**, click the "Open" button in the Home workspace, ***or*** select "File| Open" from the Menu Bar, locate and select the file, and click "Open."
2. **To open a recent file**, click the thumbnail in the Home workspace. ***Or***, select "File| Open Recent" from the Menu Bar and select the file from the list.
3. **To close a file**, select "File| Close" from the Menu Bar. **To close all open files**, select "File| Close All" instead. Photoshop will prompt you to save any un-saved changes.

Creating New Files

PSD (Photoshop Document) is the default format that Photoshop uses to save data. A .PSD file is a layered image file, preserving different elements of an image in their own layers for non-destructive editing. When an image is opened in Photoshop, the image becomes the background layer. Many edits create new layers which can be adjusted without affecting the other layers. New documents are created as .PSD files, where the background is a blank canvas.

1. Click the "Create new" button in the Home workspace, ***or*** select "File| New" from the Menu Bar.
2. **To use a preset**, select a preset category at the top of the "New Document" window ("Photo," "Print," etc.) and select the preset in the window below.
3. **To create a document with custom settings, *or* to adjust the settings of a preset,** make desired changes in the "Preset Details" panel on the right:

Width and **Height:** Sets canvas size.
Orientation: Portrait or Landscape canvas.
Resolution: 72 pixels/inch=screen content. 300=print.
Color Mode: RGB for screen display. CMYK for print.
Background Contents: Set background color or set to transparency.

4. Enter a document name (just above "Width").
5. Click the "Create" button.

Using Artboards

Adapt web and app designs for different devices and screen sizes. Artboards enable the creation of multiple versions of a design within a single document file. Artboards show as grouped layers in the Layers Panel.

1. **To create a new artboard document**, select "Artboards" in the "New Document" window when creating a new file.
2. **To create a new artboard**, select "Layer| New| Artboard" from the Menu Bar, enter a name, select an artboard preset, enter size, and click "OK."
3. **To expand/collapse the artboard's layer group**, click the arrow.
4. **To delete an artboard**, select it in the Layers Panel, click the "Delete" button in the panel.

Saving Files

To overwrite a file that has already been saved, select "File| Save" from the Menu Bar.

To save a file for the first time or to save it as a different file type, or with a different name:

1. Select "File| Save As..." from the Menu Bar.
2. Navigate to the location where you want to save.
3. Type a name in the "File name" text box.
4. Select a file type from the "Save as type" (Windows) or "Format" (Mac) drop-down (Note: .PSD files must be opened with an application such as Photoshop).
5. Click the "Save" button.

Photoshop Basics

Using the Tools Panel

The Tools Panel is a floating toolbar that hovers over the workspace at the left, by default. It contains the tools to select and edit images. When a tool is selected, its associated options appear in the Options Bar. Most buttons activate more than one tool, noted by a small white triangle in the lower-right corner of the button. The button displays the tool most recently used from its available tools.

1. **To activate a visible tool**, click it in the Tools Panel.
2. **To display and select a button's other tools**, click and hold the button, then select the tool from the menu that appears.
3. Select the tool's options in the Options Bar.

Using Other Panels

Panels allow you to select which layer of a file to edit, make adjustments, select colors, and other activities.

1. **To hide or display a panel**, ... the Menu ... e) the panel
2. **To collap**... double ar... r.
3. **To move** ...
4. **To reset a** ... Workspac...

Adjusting Image Magnification

1. Click the "Zoom" tool in the Tools Panel.
2. **To zoom in**, click the image. **To zoom out**, hold down the "Alt" key (Windows) or the "Option" key (Mac) while clicking.
3. **To quickly zoom to 100%**, select "View| 100%" from the Menu Bar. ***Or***, double-click the "Zoom" tool in the Tools Panel.
4. **To display the entire canvas within the window,** select "View| Fit on Screen" from the Menu Bar. ***Or***, double-click the "Hand" tool in the Tools Panel.
5. **To zoom to a specific magnification**, enter a value at the far-left end of the Status Bar, and press the "Enter" (Windows) or "Return" (Mac) key.

Moving the Canvas

Use the "Hand" tool to move the canvas within the window when the entire image cannot be seen.

1. Click the "Hand" tool in the Tools Panel. **To temporarily switch to the Hand tool if another tool is selected**, hold down the spacebar.
2. **To move the canvas**, click and drag it.

Moving Part of an Image

1. **To move a selected part of an image**, make a selection using one of the selection tools (see "Making Selections").
2. Click the "Move" tool in the Tools Panel.
3. Click and drag the element or selection.

Undoing Mistakes

To undo a single mistake *before* saving, select "Edit| Undo (Action name)" from the Menu Bar. **To step back through a series of steps**, repeat as desired. **To redo a single action**, select "Edit| Redo (Action name)." **To step forward through steps**, repeat as desired. **To toggle between the last two states**, select "Edit| Toggle Last State" from the Menu Bar. **To revert to the last saved version of a file**, select "File| Revert" from the Menu Bar.

To undo actions with the History Panel:

1. **To display the History Panel**, select "Window| History" from the Menu Bar. Actions are listed in the panel with the most recent action at the bottom.
2. Scroll to the desired action in the list and click it.
3. **To permanently delete an action**, select it, click the "Delete" button, and choose "Yes" to confirm.
4. **To partially undo or redo**, click the box to the left of the desired history level. The "History Brush" icon appears in the box. Click the "History Brush" tool in the Tools Panel and make brush selections. Click and drag on the part of the image to change.
5. **To create a snapshot of the image in its current state**, select the action and click the "Create a new snapshot" button. The snapshot appears the panel.

TeachUcomp, Inc.®
Computer Software Training & Continuing Education
www.teachucomp.com
info@teachucomp.com
877.925.8080

Adding Images to a File

1. **To copy an image into a .PSD file**, select "File| Place Embedded" from the Menu Bar. **To add an image as a link to the original image file**, select "File| Place Linked" from the Menu Bar.
2. Navigate to the file location and select it.
3. Click the "Place" button.

Using Rulers, Guides, and Grids

Guides and grids are horizontal and vertical lines used to precisely align and arrange objects in projects. "Smart Guides" appear automatically when you draw a shape, make a selection, or move an object. Guides are not printed or exported.

1. **To toggle the display of the rulers on or off**, select "View| Rulers" from the Menu Bar.
2. **To add a guide**, click into one of the rulers, and drag and release the guide in the desired location.
3. **To add a guide in a specific location**, select "View| New Guide" from the Menu Bar, select an orientation, enter a "Position," and click "OK."
4. **To remove a guide**, click and drag it back to the ruler. **To remove all guides**, select "View| Clear Guides" from the Menu Bar.
5. **To toggle the display of a grid, guides, or Smart Guides on or off,** select "View| Show" from the Menu Bar and then check or un-check the item.

Color Basics

Setting Foreground and Background Colors

Photoshop uses the foreground color to paint, create text, stroke and fill image selections. It uses the background color to make gradient fills and fill-in the erased areas of images. The default foreground color is black and the default background color is white. The current selections appear near the bottom of the Tools Panel and can easily be switched.

1 Default Colors 3 Foreground
2 Switch Colors 4 Background

To switch back to the default foreground color (black) and default background color (white), click the "Default Colors" button near the bottom of the Tools Panel. **To swap the current foreground and background colors**, click the "Switch Colors" button.

To set colors using the Color Picker:

1. **To view the Color Picker**, click the foreground or background color at the bottom of the Tools Panel.
2. Enter the specific color code or color values (HSB, RGB, LAB, or CMYK), if available. ***Or***, to select from the color libraries, click the "Color Libraries" button, make selections, and click "OK." ***Or***, drag the slider (small triangle) on the vertical color spectrum to select a desired hue, then click and drag in the Color Field box to select the desired color. **To restrict color choices to web-safe colors**, check the "Only Web Colors" checkbox.
3. Click "OK."

Setting Color with the Eyedropper

The Eyedropper Tool selects color from an image displayed in Photoshop or anywhere on screen.

1. Click the "Eyedropper" tool in the Tools Panel.
2. In the Options Bar, select the size of the point sample that the Eyedropper Tool uses from the "Sample Size" drop-down and the layers from which to take the sample from the "Sample" drop-down.
3. **To set the foreground color**, click anywhere on an image in Photoshop ***or*** click and drag outside of Photoshop. **To set the background color**, hold down the "Alt" key ("Option" on Mac) while clicking.

Changing an Image's Color Mode

Example: Convert an image from RGB mode used for web display to CMYK used for commercial printing.

1. Select "Image| Mode" from the Menu Bar.
2. Choose the new mode from the list.

Layers

Understanding Layers

Image content is stored in layers. Layers are like transparent sheets on which you place content. They are transparent where no layer content exists, showing other content below the layers. Layers are stacked on top of one another to separate image content. You **must** select a layer to edit its contents. You can change selected layer content without changing other layer content. Some actions create new layers (e.g. adding text, shapes, etc.). The default .PSD file format saves layers for future editing in Photoshop. Layers are managed using the Layers Panel.

Managing Layers

To display/hide the Layers Panel, select "Window| Layers" in the Menu Bar. **To select layers**, click to select them in the Layers Panel. **To select multiple layers**, hold down the "Shift" key (adjacent layers), or the "Ctrl" ("Command" on Mac) key (non-adjacent layers) while clicking. **To move layer position in the panel**, click and drag the layer(s) up or down, and release. **To hide or unhide layers and their contents**, click the "Visibility" icon to the left of the layer(s). **To create a new layer**, click the "Create a new layer" button at the bottom of the Layers Panel. **To rename a layer or group**, double-click the "Name" field to the right of the layer thumbnail, type the name, and press the "Enter" key ("Return" on Mac). **To duplicate a layer**, select it, select "Layer| Duplicate Layer..." from the Menu Bar, enter a name, select a document, and click "OK." **To delete a layer**, select it and click the "Delete layer" button.

Creating Layer Groups

Organize layers into layer groups. Expand or collapse layer groups to see or hide their contents. Grouped layers appear indented in the Layers Panel. Layer groups can be moved, hidden, renamed, duplicated, and deleted in the same manner as individual layers.

To create a new layer group:

1. Click the "Create a new group" button at the bottom of the Layers Panel.
2. Click and drag the layers onto the new group folder.

Or

1. Select the layers to group.
2. Select "Layer| Group Layers" from the Menu Bar. ***Or***, drag the selected layers onto the "Create a new group" button at the bottom of the Layers Panel.

To expand/collapse a layer group, click the triangle to the left of the group.

Linking Layers

Link layers into a single unit to collectively move or transform the layers (scale, rotate, skew, etc.). Linked layers appear with a link icon to the right.

1. Select the layers in the Layers Panel.
2. Click the "Link layers" button at the bottom.
3. **To unlink layers**, repeat steps above.

Locking Layers

Locking prevents changes from being made to an entire layer or parts of a layer.

1. Select the layer(s) in the Layers Panel.
2. Click one of the lock buttons at the top of the panel.

Lock transparent pixels
Lock image pixels
Lock position
Prevent auto-nesting
Lock all

To unlock a layer, click the lock icon that appears to the right of the layer in the Layers Panel.

Moving Layer Content

1. Click the "Move" tool in the Tools Panel.
2. Click and drag the content. ***Or***, for precise movement, press the arrow keys to "nudge" content one pixel at a time. **To nudge 10 pixels at a time**, hold down "Shift" while pressing the arrow keys.

Changing Layer Opacity

1. Select the layer in the Layers Panel.
2. Enter a value in the "Opacity" field at the top of the Layers Panel or click the downward-arrow and use the slider.

Manipulating Images

Changing the Canvas Size

Add or remove workspace around the document file without altering the images or elements within it.

1. Select "Image| Canvas Size" from the Menu Bar.
2. **To change the unit of measurement** (e.g. inches to pixels), select an option from the drop-down to the right of the "Width" or "Height" field.
3. **To add or subtract space relative to the size of the current canvas**, select the "Relative" checkbox and enter the desired values in the "Width" and "Height" fields. (e.g. Adding a value of "-2" in the "Height" field with "Inches" as a unit of measurement would reduce the height of the canvas by 2 inches). By default, changes to the canvas size are applied equally to all sides. **To restrict changes to a specific direction**, click the desired directional arrow under "Anchor."
4. Enter a value into the "Width" and "Height" fields.
5. Select a "Canvas extension color" from the drop-down.
6. Click "OK.

Resizing an Image

Adjusting the image size changes both the size of the canvas as well as the images and elements within it.

1. Select "Image| Image Size| from the Menu Bar.
2. **To view the image's dimensions in different values**, select from the "Dimensions" drop-down.
3. **To resize the image to a specific preset**, select an option from the "Fit To" drop-down.
4. **To change the unit of measurement** (e.g. inches to pixels), select an option from the drop-down to the right of the "Width" or "Height" field.
5. **To resize to specific dimensions**, enter values in the "Width," "Height," and "Resolution" fields.
6. **To constrain proportions** (width and height change proportionally to avoid distortion), be sure "Constrain Aspect Ratio" chain link icon is selected. **To deselect it** if chain icon is grayed out, check "Resample" first. *e.g. Constrained*
7. Click "OK."

Resizing One Element in a Document

1. Select the element's layer in the Layers Panel.
2. Select "Edit| Free Transform" from the Menu Bar.
3. **To constrain proportions** (width and height change proportionally to avoid distortion), hold down the "Shift" key and click and drag one of the four ***corner*** resizing handles that appear in the bounding box around the selected element. ***Or***, to resize to a specific percentage, enter a value in the "Width" and "Height" scale fields in the Options Bar.
4. Click the "Confirm" button in the Options Bar. ***Or***, press "Enter" (Windows) or "Return" (Mac).

Cropping an Image

1. Click the "Crop" tool in the Tools Panel.
2. Hover the mouse over any edge of the crop line that appears until the mouse pointer turns into a double-sided arrow, then click and drag.
3. **To constrain the crop to a common ratio**, click the "Ratio" drop-down in the Options Bar and make a selection. ***Or*, to enter a custom ratio**, enter values into the two aspect ratio fields to the right of the "Ratio" drop-down in the Options Bar.
4. Cropped pixels will be deleted by default. **To preserve the cropped pixels for later crop adjustment**, un-check the "Delete Cropped Pixels" checkbox in the Options Bar.
5. Click the "Confirm" button in the Options Bar. ***Or***, press "Enter" (Windows) or "Return" (Mac).

Straightening an Image

1. Click the "Crop" tool in the Tools Panel.
2. Click the "Straighten" button in the Options Bar.
3. Click and drag a straight line to make a new edge.
4. Click the "Confirm" button in the Options Bar. ***Or***, press "Enter" (Windows) or "Return" (Mac).

Flipping and Rotating

To flip or rotate an entire image:

1. Select "Image| Image Rotation" from the Menu Bar.
2. Select a preset. ***Or***, select "Arbitrary," enter a value in the "Angle" field, select "Clockwise" or "Counter Clockwise," and click "OK."

To rotate part of an image:

1. Select the layer or part of the image.
2. Select "Edit| Transform| Rotate" from the Menu Bar.
3. Hover the mouse outside the bounding box that appears, then click and drag in the desired direction. **To contain to 15 degree increments**, hold down the "Shift" key while dragging. **To move the pivot point for rotation from the default center**, check the "Toggle reference point" checkbox in the Options Bar, then click and drag the pivot point.
4. Click the "Confirm" button in the Options Bar. ***Or***, press "Enter" (Windows) or "Return" (Mac).

To flip part of an image:

1. Select the layer or part of the image.
2. Select "Edit| Transform" from the Menu Bar, and select either "Flip Horizontal" or "Flip Vertical."

Making Selections

Selection Basics

To edit an image file in Photoshop, first select the region to change. The selected area is indicated by a dashed border "selection marquee." Changes made will only affect the area inside the selection, protecting the rest of the image. After an initial selection, making another selection will typically replace it. **To "Add to selection" or "Subtract from selection,"** click the appropriate button in the Options Bar. **To move a selection**, click and drag within the selection area. Several tools make selections. Choosing the best tool depends on the image and selection goal. Some selections can be made using the Menu Bar:

1. **To select the entire image**, choose "Select| All."
2. **To deselect**, choose "Select| Deselect." **To undo**, choose "Select| Reselect."
3. **To select everything other than the current selection**, choose "Select| Inverse."

Rectangular and Elliptical Selections

1. Select either the "Rectangular Marquee" tool or the "Elliptical Marquee" tool in the Tools panel. Both tools share the same button.
2. Click and drag around the area to select.
3. **To select a perfect square or circle**, hold down the "Shift" key while clicking and dragging.

Making Freehand Selections

The "Lasso" tool is useful for selecting objects with complex, irregular edges.

1. Select the "Lasso" tool in the Tools Panel. It shares its button with the "Polygonal Lasso" and the "Magnetic Lasso" tool.
2. Zoom in for more precise selection, if desired.
3. Click and drag around the area to select. *Do not* release the mouse until the selection is complete by dragging back to the origin (starting) point. If released prior to reaching starting point, Photoshop will create a straight line back to it.

Making Polygonal Selections

Useful for selecting objects with straight edges.

1. Select the "Polygonal Lasso" tool in the Tools Panel. It shares its button with the "Lasso" tool and the "Magnetic Lasso" tool.
2. Click the starting point, then click a second point to create a straight line. Repeat as necessary. **To draw freehand**, hold down the "Alt" key while dragging.
3. Click the starting point to complete the selection.

Using the Magnetic Lasso Tool

Useful for selecting objects with complex, irregular edges set against a high-contrast background. Makes a selection by "snapping" to image edges.

1. Select the "Magnetic Lasso" tool in the Tools Panel. It shares its button with the "Lasso" tool and the "Polygonal Lasso" tool.
2. Click the starting point and drag along the edge. Photoshop adds "fastening points" along the edge as it detects it.
3. **To manually add a fastening point**, click once. **To delete a fastening point**, press the "Delete" key.
4. **To specify the lasso's sensitivity to edges**, enter a value between 1% and 100% in the "Contrast" field in the Options Bar.
5. **To specify the rate at which the lasso sets fastening points**, enter a value between 0 and 100 in the "Frequency" field in the Options Bar.
6. Click the starting point to complete the selection.

Selecting with the Brush Tool

Use the "Brush" tool in "Quick Mask Mode" to paint and create a free-form selection.

1. Choose "Select| Edit in Quick Mask Mode" from the Menu Bar.
2. Select the "Brush" tool in the Tools Panel.
3. **To change brush settings, including size**, click to open the "Brush Preset Picker" in the Options Bar and make desired modifications. "Hardness" refers to the sharpness of the edge of the selection made by the brush. A harder brush creates a sharper edge.
4. **To add to the mask**, change the brush color to black in the Tools Panel. **To subtract from the mask**, change the brush color to white.
5. Click and drag to paint the areas to be ***excluded*** from the selection. Everything else will be selected. The excluded area (mask) will be tinted red.
6. When done, choose "Select| Edit in Quick Mask Mode" from the Menu Bar. The selection marquee appears.

Using the Quick Selection Tool

Use the "Quick Selection" tool to quickly "paint" a selection based on the colors and tones of the image using an adjustable round brush tip. As you drag, the selection expands outward and automatically finds and follows defined edges within the image.

1. Select the "Quick Selection" tool in the Tools Panel. It shares a button with the "Magic Wand" tool and the "Object Selection" tool.
2. **To change brush settings including size**, click to open the "Brush Settings" in the Options Bar and make desired modifications.
3. **To include all layers in the selection**, check the "Sample All Layers" checkbox in the Options Bar. By default, only the selected layer is used.
4. Click and drag to create the selection.

Refining Selections

1. With one of the selection tools selected, click "Select and Mask" in the Options Bar. Or, choose "Select| Select and Mask" from the Menu Bar. The "Select and Mask" workspace opens. The Tools Panel is replaced with limited tools ("Quick Selection," "Refine Edge Brush," "Brush," "Object Selection,""Lasso," "Hand," and "Zoom"). The Properties Panel appears at the right of the window.
2. Use the selection tools to make a selection (if not already done).
3. Make any desired changes to edge settings in the Properties Panel. Select a view from the "View" drop-down at the top of the Properties Panel to adjust how the mask and image is displayed. Note: The "Global Refinements" settings affect the entire selection, not just part of it.
4. **To refine a selection manually**, click the "Refine Edge Brush" in the Tools Panel and click and drag along the edge of the selection.
5. Click "OK" in the Properties Panel when done.

Text

Adding Text

There are two basic types of text in Photoshop: "point type" and "paragraph type." Point type is useful for entering a word or single line of text and paragraph type is useful for entering and formatting multiple lines of text as one or more paragraphs. Paragraph text is added in a text box. Text will wrap to the next line when it reaches the end of the box. Text can be entered horizontally or vertically.

1. Select the "Horizontal Type" tool or the "Vertical Type" tool in the Tools Panel. They share the same button as the "Vertical Type Mask" tool and the "Horizontal Type Mask" tool.
2. **To insert "point type" text**, click once into the document. **To insert "paragraph type" text**, click and drag in the document to create a text box to hold the text.
3. Type the desired text.
4. **To adjust the font**, use settings in the Options Bar.

5. **To access additional formatting options**, select "Window| Character" or "Window| Paragraph" from the Menu Bar. ***Or***, click the "Toggle Character and Paragraph Panels" button in the Options Bar. Click on the "Character" and "Paragraph" panels and make selections from the various options.
6. **To increase the size of a text box** (if text does not fit into existing text box), click one of the resizing handles around the text box and drag to a new size.
7. Click the "Commit" button in the Options Bar.

Editing Existing Text

Use the type tools to edit existing text. Resizing a text box without a type tool selected may distort text. Font rasterization is the process of converting text from a vector description (as found in scalable fonts) to raster or bitmap. Rasterized text cannot be edited.

1. Select the layer that contains the text to edit.
2. Select the "Horizontal Type" tool or the "Vertical Type" tool in the Tools Panel.
3. Click into the existing text so the cursor appears.
4. Type, select, and edit, as desired.
5. Click the "Commit" button in the Options Bar.

Warping Text

1. Add or edit text normally.
2. Prior to clicking the "Commit" button, click the "Warp Text" button in the Options Bar.
3. Select a warp style from the "Style" drop-down.
4. Make any desired adjustments using the options below the "Style" drop-down. Click the "OK" button.
5. Click the "Commit" button in the Options Bar.

Rotating, Skewing, Scaling Text

1. Select the "Horizontal Type" tool or the "Vertical Type" tool in the Tools Panel. Click into the text.
2. **To rotate**, hold the mouse pointer outside of the bounding box until a curved arrow appears, then click and drag. For 15 degree increments, hold down "Shift" while dragging.
3. **To skew**, hold down the "Ctrl" key, then click and drag one of the side (non-corner) resizing handles.
4. **To scale**, hold down the "Ctrl" key, then click and drag one of the corner resizing handles.
5. Click the "Commit" button in the Options Bar.

Creating a Type Selection Mask

A type selection is a selection area that is in the shape of text that you type. You can then apply most of the same effects that you apply to normal selections.

1. Select the "Horizontal Type Mask" tool or the "Vertical Type Mask" tool in the Tools Panel.
2. **To insert "point type" text**, click once into the document. **To insert "paragraph type" text**, click and drag in the document. The artboard turns red.
3. Type the desired text.
4. Click the "Commit" button in the Options Bar.
5. Apply desired fills, colors, patterns, gradients, etc.

Enhancing and Repairing

Understanding Layer Masks

Use masks to hide portions of a layer and reveal the layers below. There are two types of masks. Layer masks are resolution-dependent bitmap images that are edited with the painting or selection tools. Vector masks are resolution independent and are created with a pen or shape tool. Masks are nondestructive, meaning they can be edited later while retaining the original pixels they hide. For example, to replace a background in an image, place the image's layer on top of the image layer containing the new background, apply a mask to the top layer, and use the brush tool to hide the pixels to reveal the new background. A layer with a mask applied displays a rectangle to the right of the layer's link icon in the Layers Panel. The white in the rectangle shows parts that are not masked. The black is masked.

Applying Layer Masks

To add a mask to any layer:

1. Select the layer in the Layers Panel.
2. Click the "Add Layer Mask" button at the bottom of the Layers Panel.

To select what is masked:

1. Select the layer in the Layers Panel.
2. Use a selection tool to select what to keep *unmasked*.
3. Click the "Add Layer Mask" button at the bottom of the Layers Panel. The mask covers unselected areas.

Or

1. *Double-click* the mask preview in the Layers Panel. If prompted, click "Enter Select and Mask." The Properties Panel appears.
2. Select a mask view from the "View" drop-down at the top of the Properties Panel.
3. **To add or remove from the mask**, select a brush or selection tool in the Tools Panel. Click the "Add to Selection" button (to unmask) or the "Subtract from Selection" button (to mask) in the Options Bar. **To examine all layers when creating the mask**, click the "Sample All Layers" checkbox in the Options Bar. Apply with the selection or brush tool.
4. **To adjust the mask**, use the settings in the Properties Panel.
5. Click "OK" in the Properties Panel when done.

Applying Layer Styles

Layer styles are effects ("Drop Shadow," "Outer Glow," "Stroke," etc.) applied to a layer or layer group. More than one style can be applied to a layer. The layer effects icon appears to the right of the layer's name in the Layers Panel when effects are applied.

1. Select the layer in the Layers Panel.
2. Click the "Add a layer style" button at the bottom of the Layers Panel. ***Or***, select "Layer| Layer Style" from the Menu Bar. Select "Blending Options" from the menu. The "Layer Style" dialog box opens.
3. Select a "Style" on the left. A checkmark appears. **To un-apply a style**, un-check the box.
4. Change style settings using controls on the right.
5. Click "OK" when done.

Note: Layer styles cannot be directly applied to a background layer. Apply a mask first, then a style.

Removing with Content-Aware Fill

Content-Aware analyzes the pixels surrounding a selection and fills the selection with a blend of those pixels. Best for images with a consistent background.

1. Select the content to remove.
2. Select "Edit| Fill" from the Menu Bar.
3. Select "Content-Aware" from the "Contents" drop-down.
4. Adjust settings as desired.
5. Click "OK."

Healing Imperfections

The "Healing Brush" tool corrects imperfections in an image by blending them with the pixels sampled from the surrounding image. The "Spot Healing Brush" tool works well for small areas and samples pixels from the area around the area to be retouched.

1. Click the "Healing Brush" tool or the "Spot Healing Brush" tool in the Tools Panel. They share a button in the Tools Panel.
2. Set brush size and other settings in the Options Bar.
3. If using the Healing Brush tool, hold down "Alt" ("Option" for Mac) and click on a source area from which to sample.
4. Click and drag on the area to heal.

Cloning Content

The "Clone Stamp" tool duplicates part of an image by sampling "source" pixels and then painting them into another part of the image.

1. Select the "Clone Stamp" tool in the Tools Panel. It shares a button with the "Pattern Stamp" tool.
2. **To change brush settings, including size**, click to open the "Brush Preset Picker" in the Options Bar and make desired modifications.
3. **To select pixels in layers other than the current layer**, select "All Layers" from the "Sample" drop-down in the Options Bar.
4. Move the mouse pointer over the area to copy. Hold down "Alt" ("Option" for Mac) and click.
5. **To clone the pixels from the source**, click and drag in another part of the image. The source will move as the mouse is dragged (indicated by a cross).

Applying Adjustments and Fills

Adjustments such as "Exposure," "Brightness/Contrast," and "Black & White" are layers that apply color and tonal changes to images without permanently altering the original pixels. Fill layers are filled with colors, patterns, or gradients. Adding an adjustment or fill layer automatically creates a new layer with a linked mask.

1. Select "Layer| New Adjustment Layer" or "Layer| New Fill Layer" from the Menu Bar. ***Or***, click the "Create new fill or adjustment layer" button at the bottom of the Layers Panel.
2. Select an Adjustment or Fill type from the menu. A new layer is created and the Properties Panel opens. For fill layers, the "Color Picker," "Gradient Fill," or "Pattern Fill" dialog box also opens.
3. To limit adjustments to unmasked parts of the image, set masking (see ***Applying Layer Masks***).
4. Make desired selections in the dialog box and/or Properties Panel.

Adjusting Exposure

Use the "Dodge" tool to lighten parts of an image. Use the "Burn" tool to darken.

1. Click the "Dodge" tool or the "Burn" tool in the Tools Panel. They share a button in the panel.
2. **To change brush settings, including size**, click to open the "Brush Preset Picker" in the Options Bar and make desired modifications.
3. **To set the strength of the effect**, set the "Exposure" drop-down slider in the Options Bar.
4. Click and drag in the image. **To compound the effect**, click and drag over the same area again.

Adjusting Color Saturation

Use the "Sponge" tool to saturate or desaturate colors.

1. Click the "Sponge" tool in the Tools Panel. It shares a button with the "Dodge" and "Burn" tools.
2. **To change brush settings, including size**, click to open the "Brush Preset Picker" in the Options Bar and make desired modifications.
3. Select "Saturate" or "Desaturate" from the "Mode" drop-down in the Options Bar.
4. **To set the strength of the effect**, set the "Flow" drop-down slider in the Options Bar.
5. Click and drag in the image. **To compound the effect**, click and drag over the same area again.

Blurring and Sharpening an Image

The "Blur" tool is useful for softening hard lines. The "Sharpen" tool sharpens the edges.

1. Click the "Blur" tool or the "Sharpen" tool in the Tools Panel. They share a button with the "Smudge" tool.
2. **To change brush settings, including size**, click to open the "Brush Preset Picker" in the Options Bar and make desired modifications.
3. **To set the strength of the effect**, set the "Strength" drop-down slider in the Options Bar.
4. Click and drag in the image. **To compound the effect**, click and drag over the same area again.

Keyboard Shortcuts

Tools — Win/Mac

Press the shortcut key to activate the tool. For tools that share a button, hold down the "Shift" key while pressing the letter to rotate through the tools.

Tools	Win/Mac
Move/Artboard	V
Rectangular/Elliptical Marquee	M
Lasso/Polygonal Lasso/Magnetic Lasso	L
Object Selection/Quick Selection/Magic Wand	W
Crop/Perspective Crop/Slice/Slice Select	C
Frame	K
Eyedroppers/Color Sampler/Ruler/Note/Count	I
Healing Brushes/Patch/Content Aware/Red Eye	J
Brush/Pencil/Color Replacement/Mixer Brush	B
Clone Stamp/Pattern Stamp	S
History Brush/Art History Brush	Y
Eraser/Background Eraser/Magic Eraser	E
Gradient/Paint Bucket/3D Material Drop	G
Dodge/Burn/Sponge	O
Pen/Freeform Pen/Curvature Pen	P
Text Tools	T
Path Selection/Direct Selection	A
Shapes	U
Hand	H
Rotate View	R
Zoom	Z
Set Default Foreground/Background Colors	D
Switch Foreground/Background Colors	X
Quick Mask Mode Edit	Q

Brushes	Win	Mac
Increase Brush Size	]	]
Decrease Brush Size	[	[
Increase Brush Hardness	Shift+]	Shift+]
Decrease Brush Hardness	Shift+[	Shift+[

Documents	Win	Mac
Change Canvas Size	**Alt+Ctrl+C**	**Option+⌘+C**
Change Image Size	**Alt+Ctrl+I**	**Option+⌘+I**
Close Document	**Ctrl+W**	**⌘+W**
Deselect All	**Ctrl+D**	**⌘+D**
Invert Selection	**Shift+Ctrl+I**	**Shift+⌘+I**
New Document	**Ctrl+N**	**⌘+N**
Open Document	**Ctrl+O**	**⌘+O**
Print	**Ctrl+P**	**⌘+P**
Redo	**Ctrl+Z**	**⌘+Z**
Save Document	**Ctrl+S**	**⌘+S**
Save Copy of Document	**Shift+Ctrl+S**	**Shift+⌘+S**
Select All	**Ctrl+A**	**⌘+A**
Select and Mask	**Alt+Ctrl+R**	**Option+⌘+R**
Step Backward	**Alt+Ctrl+Z**	**Option+⌘+Z**
Step Forward	**Shift+Ctrl+Z**	**Shift+⌘+Z**
Undo	**Ctrl+Z**	**⌘+Z**
Zoom In	**Ctrl+*plus sign***	**⌘+*plus sign***
Zoom Out	**Ctrl+*minus sign***	**⌘+*minus sign***

ISBN 978-1-941854-72-3

9 781941 854723

Made in the USA